Math Interview Questions

Jean Peyre, Math Interview Questions.

Paperback Edition September 2020
For any question, remark or typo please email admin@editionsducourt.com

Copyright © 2020 by the author. All rights reserved.

Math Interview Questions

Jean Peyre

ÉDITIONS DUCOURT

Contents

Brainteasers 3
Brainteasers - Solutions 15
Math Cheatsheet 51
A Humble Request 55
   Index 57

Chapter 1

Brainteasers

Brainteasers

Difficulty: ♠ Medium ♠♠ Hard ♠♠♠ Very Hard

1.1 Triangle Impossible I ♠

We randomly break a stick of length 1 into three pieces, What is the probability that the pieces can form a triangle? (breaking points are uniformly distributed between 0 and 1).

Solution in page 15

1.2 Triangle Impossible II ♠♠

Let A, B and C three independent random variables uniformly distributed between 0 and 1. We make 3 sticks respectively of length A, B and C. What is the probability that the sticks can form a triangle?

Solution in page 16

1.3 A Prime Number ♠

We consider a prime number $p \geq 5$. Prove that 24 divides $(p^2 - 1)$ i.e. $24|(p^2 - 1)$.

Solution in page 18

1.4 The Last Prime ♠

Prove that there is an infinity of prime numbers.

Solution in page 18

1.5 Erdős Subsequences ♠♠♠

We consider a sequence u_n, $n \in [1, 300]$ composed of distinct real numbers. Show that we can extract a strictly increasing or strictly decreasing subsequence $u_{\phi(n)}$ containing at least 17 elements.

Solution in page 18

1.6 Omelette ♠♠

You are given two eggs, and access to a 100-storey building. Both eggs are identical. The aim is to find out the highest floor from which an egg will not break when dropped out of a window from that floor. If an egg is dropped and does not break, it is undamaged and can be dropped again. However, once an egg is broken, that's it for that egg. If an egg breaks when dropped from floor n, then it would also have broken from any floor above that. If an egg survives a fall, then it will survive any fall shorter than that.

What strategy should you adopt to minimize the number of egg drops it takes to find the solution?

Solution in page 19

1.7 A Hard Pill to Swallow ♠

A blind man is alone on a deserted island. He has two blue pills and two red pills. He must take exactly one red pill and one blue pill to survive. How does he do it?

Solution in page 20

1.8 Game Theory ♠♠♠

Player A invites player B to play the following game: A picks an integer n between 1 and 100, and writes it on a paper. B tries to guess n. If he succeeds, he receives n dollars. What is the fair price of the game, and what should be the strategy of B?

Solution in page 21

1.9 Stable Equilibrium I ♠

N tigers circle around an antelope. If a tiger eats an antelope or another tiger, it falls asleep and it becomes a potential meal for the remaining tigers. Tigers will eat if it does not endanger their life. The antelope keeps grazing quietly. Why?

Solution in page 21

1.10 Stable Equilibrium II ♠♠

100 silent monks live in a monastery with no mirrors or reflective surfaces and one important rule: no red eyes! If a monk discovers he has red eyes he commits suicide at midnight. They live happily together in peace until a tourist visiting the monastery says "at least one of you has red eyes!". What happens next?

Solution in page 22

1.11 Need For Speed ♠

A car travels 100km in 1 hour. Show that, at some point, its speed was exactly 100km/h.

Solution in page 23

1.12 Russian Coin I ♠

Three players A, B and C sit around a table. They have a fair coin which gives heads or tails with a probability $\frac{1}{2}$. Player A tosses the coin, if he gets heads he wins, and the game is over. Otherwise he gives the coin to B, who is sitting at his right hand side. If B gets heads he wins, otherwise he gives the coin to C etc... What is the probability for each player to win the game?

Solution in page 23

1.13 Russian Coin II ♠

Three players A, B and C sit around a table. They have a strange coin which gives heads or tails with a probability $\frac{1}{4}$, and stays stuck on its side with a probability $\frac{1}{2}$. Player A tosses the coin, if he gets side he wins, and the game is over. Otherwise if A gets heads he gives the coin to B, who is sitting at his right hand side. If A gets tails he gives the coin to C, who is sitting at his left hand side. The next player restarts the same process. What is the probability for each player to win the game?

Solution in page 24

1.14 Be My Guest ♠♠

N guests are queuing at the entrance to get seated at a wedding table. Every guest has an assigned seat number but the first guest to choose his seat is too drunk and takes a random seat. The remaining guests choose their seat according to the following rule:

- if their assigned seat is available they take it

- if their assigned seat is taken they choose randomly an available seat

What is the probability that the last person gets his assigned seat?

Solution in page 25

1.15   4 Coins, 1 Table ♠♠♠

4 coins are placed at the corners of a rotating table and the player is blindfolded. At every turn, the player can flip as many coins as he wants, and ask the game master if the coins are all showing heads. If they are all heads, the players wins, otherwise the game master can arbitrarily rotate the table before the next turn. Is there a winning strategy for the player?

Solution in page 26

1.16   N Coins, 1 Table ♠♠♠

2 players take turns placing coins on a large perfectly round table. Coins can not overlap and all the coin surface must be in contact with the table. The first player who can't place a coin loses. Is it better to play first and is there a winning strategy?

Solution in page 27

1.17   Regression Mirror ♠

Suppose that X and Y are mean zero, unit variance random variables. If least squares regression (without intercept) of Y against X gives a slope of β (i.e. it minimises $\mathbb{E}[(Y - \beta X)^2]$), what is the slope of the regression of X against Y?

Solution in page 28

1.18   Bayesian Kids ♠

I meet someone with 2 children, and I learn that one of the children is a boy. What's the probability that the other child is also a boy? What if one of the children is a boy born on a Tuesday?

Solution in page 28

1.19 The Last Digit ♠

Consider all 100 digit numbers, i.e. those between 0 to $(10^{100} - 1)$, inclusive. For each number, take the product of non-zero digits (treat the product of digits of 0 as 1), and sum across all the numbers. What's the last digit?

Solution in page 29

1.20 Repeated Contraction ♠

Let $R(n)$ be a random draw of integers between 0 and $n-1$ (inclusive). I repeatedly apply R, starting at 10^{100}. What's the expected number of repeated applications until I get zero?

Solution in page 30

1.21 Domino's Pizza ♠♠

How many ways are there to tile dominos (with size 2×1) on a grid of $2 \times n$? How about on a grid of $3 \times 2n$?

Solution in page 30

1.22 Nash's Car ♠♠♠

A company has a competition to win a car. Each contestant needs to pick a positive integer. If there's at least one unique choice, the person who made the smallest unique choice wins the car. If there are no unique choices, the company keeps the car and there's no repeat of the competition. It turns out that there are only three contestants, and you're one of them. Everyone knows before picking their numbers that there are only three contestants. How should you make your choice?

Solution in page 33

1.23 Correlation Impossible I ♠

If X, Y and Z are three random variables such that X and Y have a correlation of 0.9, and Y and Z have correlation of 0.8, what are the minimum and maximum correlation that X and Z can have?

Solution in page 35

1.24 Correlation Impossible II ♠♠

If $X_1, X_2...X_n$ are n random variables such that

$$\mathrm{Corr}(X_i, X_j) = \rho \quad \text{for } i \neq j$$

what are the minimum and maximum values that ρ can have?

Solution in page 35

1.25 The Dark Side of the Die ♠

How many times do I have to roll a die until all six sides appear?

Solution in page 36

1.26 Bonus Day ♠

Five pirates P_i have 100 gold coins. They have to divide up the loot. In order of seniority (suppose pirate P_5 is most senior, P_1 is least senior), the most senior pirate proposes a distribution of the loot. They vote and if at least 50% accept the proposal, the loot is divided as proposed. Otherwise the most senior pirate is executed, and they start over again with the next senior pirate. Which solution does the most senior pirate propose? Assume they are very intelligent and extremely greedy (and that they would prefer not to die).

Solution in page 37

1.27 Secret Polynomial ♠♠

We consider a polynomial $P(x)$ which all coefficients are positive ($a_i \geq 0$). The polynomial is in a black box and we can only retrieve its value in given points. In how many points do we need to value the polynomial in order to find the values of all the coefficients?

Solution in page 38

1.28 Drunk Mutant Ninja Ant ♠♠

An ant starts a walk from a cube vertex, it walks on the edges and at every vertex it chooses to walk one of the available edges (including the edge it came from) with an equal probability. How many edges will the ant cross in average to come back to the starting point?

Solution in page 38

1.29 Dog Day Afternoon ♠♠

You are standing at the centre of a circular field of radius R. The field has a low wire fence around it. Attached to the wire fence (and restricted to running around the perimeter) is a large, sharp-fanged, hungry dog. You can run at speed v, while the dog can run four times as fast. What is your running strategy to escape the field?

Solution in page 39

1.30 Exp Pi Ring ♠♠

Is $\pi^e > e^\pi$?

Solution in page 40

1.31 Brilliant ♠

You're in a room with three light switches, each of which controls one of three light bulbs in the next room. You need to determine which switch controls which bulb. All lights are off to begin, and you can't see into one room from the other. You can inspect the other room only once. How can you find out which switches are connected to which bulbs?

Solution in page 41

1.32 Lognormal Expectation ♠

Calculate $\mathbb{E}\left(\exp(X)\right)$ when is X is a normally distributed random variable

$$X \sim \mathcal{N}\left(\mu, \sigma^2\right)$$

Solution in page 41

1.33 Cumulative Brownian ♠

Calculate $\mathbb{E}(\Phi(B_t))$ where B_t a brownian motion and Φ the standard normal cumulative distribution.

Solution in page 42

1.34 Learn the Ropes ♠

You have two ropes coated in an oil to help them burn. Each rope will take exactly 1 hour to burn all the way through. However, the ropes do not burn at constant rates, there are spots where they burn a little faster and spots where they burn a little slower, but it always takes 1 hour to finish the job.

With a lighter to ignite the ropes, how can you measure exactly 90 minutes? And how can you measure 45 minutes?

Solution in page 42

1.35 Bayes Bias ♠

I have one fair coin and one biased two headed coin, and I put both in my pocket. I randomly choose one coin and flip it. It shows heads. What is the probability that the coin has Tails on the other side? What if you flip it n times and get heads n times?

Solution in page 43

1.36 Die Hard ♠

You've got to defuse a bomb by placing exactly 4 gallons of water on a sensor. You only have a 5 gallon jug and a 3 gallons jug on hand. How do you proceed?

Solution in page 44

1.37 Blind Coins ♠♠

You have a 100 coins laying flat on a table, each with a head side and a tail side. 10 of them are heads up, 90 are tails up. You can't feel, see or in any other way find out which 10 are heads up. How can you split the coins into two piles so there are the same number of heads-up coins in each pile?

Solution in page 44

1.38 Russian Dilemma ♠

We are to play a version of Russian Roulette, the revolver is a standard six shooter but I will put two bullets in the gun in consecutive chambers. I spin the chambers, put the gun to my head pull the trigger and survive. I hand you the gun and give you a choice... You may put the gun straight to your head and pull the trigger, or you may re-spin the gun before you do the same.

What is your choice and why? How does this differ from the case with only one bullet?

Solution in page 45

1.39 Rain Check ♠♠

You're about to get on a plane to Seattle. You want to know if it's raining there. You call 3 random friends who live there and ask each if it's raining. Each friend has a 2/3 chance of telling you the truth and a 1/3 chance of messing with you by lying. All 3 friends tell you that "Yes" it is raining. What is the probability that it's actually raining in Seattle?

Solution in page 45

1.40 No Time to Die ♠

A regular clock has an hour and minute hand. At 12 midnight the hands are exactly aligned. When is the next time they will exactly align or overlap?

Solution in page 46

Chapter 2

Brainteasers - Solutions

Brainteasers - Solutions

2.1 Triangle Impossible I - Solution

Question : We randomly break a stick of length 1 into three pieces, What is the probability that the pieces can form a triangle? (breaking points are uniformly distributed between 0 and 1).

Solution : An elegant and effective method to solve the problem is to visualize it geometrically. Let us define x and y the two breaking points and let us assume $x \geq y$ (the case $x \leq y$ is symmetric). In order to form a triangle we need the longest piece to be shorter than the sum of the two other pieces.

$$\text{In the case } x \geq y \text{ we need } \begin{cases} x \geq \frac{1}{2} \\ y \leq \frac{1}{2} \\ (x-y) \leq \frac{1}{2} \end{cases} \qquad (1)$$

We translate these conditions in the chart below, the chart on the left side is for the case $x \geq y$ and the chart on the right side is for the general case. The grey area represents the cases where we can form a triangle, we see that in both charts the grey area is a quarter of the full area. The probability to form a triangle is therefore $\frac{1}{4}$.

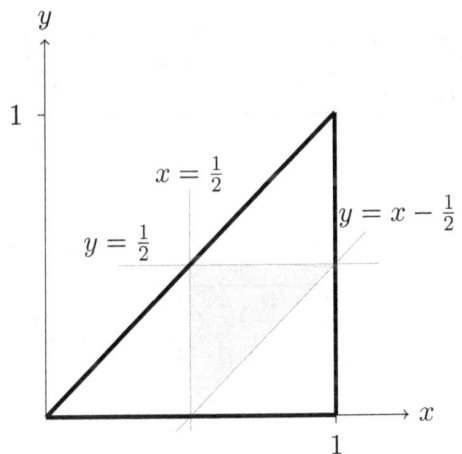
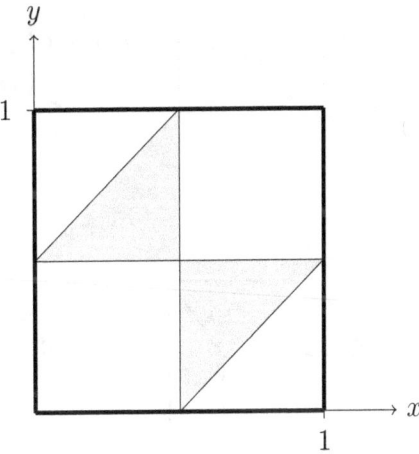

We can also answer using integrals. We consider here that $x \leq \frac{1}{2}$ (the other case is symmetric), and for $x \in [0, \frac{1}{2}]$ we see that y must be greater than $\frac{1}{2}$ and lower than $(x + \frac{1}{2})$.

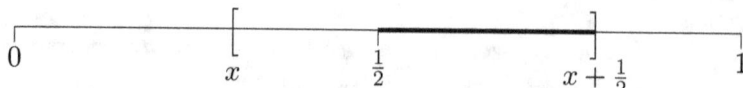

Translated into integrals, and using the symmetry argument we have P the probability to form a triangle

$$P = 2 \int_0^{\frac{1}{2}} x\,dx = 2 \left[\frac{x^2}{2}\right]_0^{\frac{1}{2}} = \frac{1}{4}$$

2.2 Triangle Impossible II - Solution

Question : Let A, B and C three independent random variables uniformly distributed between 0 and 1. We make 3 sticks respectively of length A, B and C. What is the probability that the sticks can form a triangle?

Solution : This problem can be solved elegantly with a drawing. We work this time on a cube and the condition to form a triangle is that no stick is longer that the sum of the others.

$$\text{we need } \begin{cases} A \leq (B+C) \\ B \leq (A+C) \\ C \leq (B+A) \end{cases} \qquad (2)$$

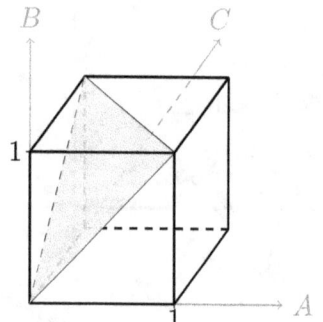
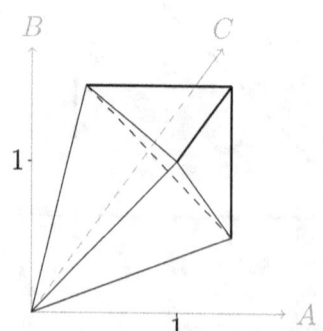

The valid volume is the diamond on the right. To calculate this volume we subtract 3 times the volume of the pyramid below from the the original cube.

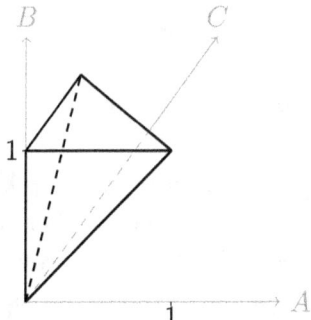

The volume of this pyramid is $V_{pyramid} = base.height.\frac{1}{3} = \frac{1}{2}.1.\frac{1}{3} = \frac{1}{6}$. Therefore $V_{diamond} = 1 - 3.\frac{3}{6} = \frac{1}{2}$. The probability we are looking for is $P = \frac{1}{2}$.

The question can also be solved with integrals

$$I = \int_{x=0}^{x=1} \int_{y=x}^{y=1} \int_{z=y}^{1 \wedge (x+y)} dxdydz$$

$$I = \int_{x=0}^{x=1} \int_{y=x}^{y=1} \left[x \wedge (1-y)\right] dxdy$$

$$I = \int_{x=0}^{x=1} \int_{Y=0}^{Y=1-x} \left[x \wedge Y\right] dxdY; \quad Y = (1-y)$$

We decompose I into 3 terms

$$I = I_1 + I_2 + I_3$$

$$I = \int_{x=0}^{x=\frac{1}{2}} \int_{Y=0}^{Y=x} YdxdY + \int_{x=0}^{x=\frac{1}{2}} \int_{Y=x}^{Y=1-x} xdxdY + \int_{x=\frac{1}{2}}^{x=1} \int_{Y=0}^{Y=1-x} YdxdY$$

$$I_1 = \int_{x=0}^{x=\frac{1}{2}} \int_{Y=0}^{Y=x} YdxdY = \int_{x=0}^{x=\frac{1}{2}} \frac{x^2}{2} dx$$

$$I_2 = \int_{x=0}^{x=\frac{1}{2}} \int_{Y=x}^{Y=1-x} xdxdY = \int_{x=0}^{x=\frac{1}{2}} x(1-2x)dx$$

$$I_3 = \int_{x=\frac{1}{2}}^{x=1} \int_{Y=0}^{Y=1-x} YdxdY = \int_{x=\frac{1}{2}}^{x=1} \frac{(1-x)^2}{2} dx = \int_{s=\frac{1}{2}}^{s=0} -\frac{s^2}{2} ds = \int_{x=0}^{x=\frac{1}{2}} \frac{s^2}{2} dx$$

$$I = \int_{x=0}^{x=\frac{1}{2}} \frac{x^2}{2} + \frac{x^2}{2} + x(1-2x)dx = \int_{x=0}^{x=\frac{1}{2}} x(1-x)dx = \left[\frac{x^2}{2} - \frac{x^3}{3}\right]_0^{\frac{1}{2}} = \frac{1}{12}$$

$$P = 6I = \frac{1}{2}$$

2.3 A Prime Number - Solution

Question : We consider a prime number $p \geq 5$. Prove that 24 divides $(p^2 - 1)$ i.e. $24 | (p^2 - 1)$.

Solution : In order to prove that $24|(p^2-1)$ we can prove that $8|(p^2-1)$ and $3|(p^2-1)$.

We note first that $p^2 - 1 = (p-1)(p+1)$, and p being a prime number, $(p-1)$ and $(p+1)$ are two consecutive even integers. Therefore both can be divided by 2 and one of them can be divided by 4. We have proved that $8|(p^2-1)$.

We also observe that $p-1$, p and $p+1$ are 3 consecutive integers. One of them is necessarily divisible by 3 and it can not be p because it is a prime number. This shows that $3|(p^2-1)$ and therefore $24|(p^2-1)$.

2.4 The Last Prime - Solution

Question : Prove that there is an infinity of prime numbers.

Solution : This is a classic proof in number theory. We proceed by contradiction, we assume that the set of all prime numbers is finite $\{n_0, n_1, ..., n_M\}$. We consider now the integer

$$K = 1 + \prod_{i=0}^{M} n_i$$

K can not be divided by any of the prime numbers in our set because for all of them we have K modulo n_i equal to 1 i.e. $K[n_i] = 1$. This means that K is a prime number which was not in our set, hence the contradiction.

2.5 Erdős Subsequences - Solution

Question : We consider a sequence u_n, $n \in [1, 300]$ composed of distinct real numbers. Show that we can extract a strictly increasing or strictly decreasing subsequence $u_{\phi(n)}$ containing at least 17 elements.

Solution : The question is about the longest monotonous subsequence that can be extracted from a given sequence. Intuitively this length increases with the length of the original sequence. We denote I_i (resp. D_i) the longest increasing (resp. decreasing) subsequence which last element is u_i. The application $i \mapsto \{I_i, D_i\}$ is injective

$$m < n \Rightarrow \begin{cases} u_m < u_n; I_n > I_m \\ \text{or} \\ u_m > u_n; D_n > D_m \end{cases}$$

Therefore once $n > p^2$ at best we can fill the square $[1,p] \times [1,p]$ and be guaranteed to find a monotonous subsequence of length $p+1$. Actually for any n we can find a monotonous subsequence of length $\lceil \sqrt{n} \rceil$. In our case $n = 300$ and we can extract 17 ordered elements.

So why is Erdos in the title? Because the Erdős–Szekeres theorem guarantees that any sequence of distinct real numbers with length at least $(r-1)(s-1)+1$ contains a monotonically increasing subsequence of length r or a monotonically decreasing subsequence of length s. In this case $r = s = 17$ and 290 is required number in the sequence.

2.6 Omelette - Solution

Question : You are given two eggs, and access to a 100-storey building. Both eggs are identical. The aim is to find out the highest floor from which an egg will not break when dropped out of a window from that floor. If an egg is dropped and does not break, it is undamaged and can be dropped again. However, once an egg is broken, that's it for that egg. If an egg breaks when dropped from floor n, then it would also have broken from any floor above that. If an egg survives a fall, then it will survive any fall shorter than that.

What strategy should you adopt to minimize the number of egg drops it takes to find the solution?

Solution : The objective is to minimize the number of attempts in the worst case. If we had only one egg to solve the problem we would have needed to start at the first floor and to go up one floor for every new attempt. In the worst case we would have needed 100 attempts. If we have 2 eggs we can improve this strategy and skip floors when using the first egg. If the first egg breaks we can single out an interval of floors. We can then use the second egg to test the floors in the interval one by one from the bottom. The crucial question is the choice of intervals to skip with the first egg. We denote u_i the sequence of floors from which the first egg is thrown and $W(k)$ the number of attempts needed in the worst case to solve the problem with 2 eggs and k floors. After the first attempt at u_1, if the egg breaks we need try all the floors between 1 and $u_i - 1$. Otherwise we still have 2 eggs and

$(100 - u_i)$ remaining floors to test

$$W(100) = \max\left(u_1, 1 + W(100 - u_1)\right)$$

we repeat the same reasoning until the ith floor

$$W(100) = \max\left(u_1, u_2 - u_1 + 1, u_3 - u_2 + 2, \ldots, 1 + W(100 - u_i)\right)$$

we denote v_i the increments sequence $v_i = u_i - u_{i-1}$, and $v_1 = u_1$. The equation becomes

$$W(100) = \max\left(v_1, v2 + 1, v_3 + 3, \ldots, 1 + W\left(100 - \sum_{k=1}^{i} v_k\right)\right)$$

and the full formula for the number of attempts is

$$W(100) = \max\left(v_1, v2 + 1, v_3 + 2, \ldots, v_n + n - 1\right)$$

we minimize this maximum when all the arguments are equal. On the other hand the increments sum to 100

$$\sum_{k=1}^{n} v_k = 100$$

We denote $M = v_i + i - 1$, and for a given n the condition on M is

$$\sum_{k=1}^{n}(M - k + 1) = nM + n - \sum_{k=1}^{n} k > 100$$

$$M > \frac{100}{n} - 1 + \frac{n+1}{2}$$

we calculate the derivative of the right side to find that the minimum is reached for $n = \sqrt{200} \approx 14.14$ and therefore $M = 15$. In the worst case we will need 15 attempts and the sequence of floors to test with the first egg is

$$15, 29, 42, 54, 65, 75, 84, 92, 99, 100$$

2.7 A Hard Pill to Swallow - Solution

Question : A blind man is alone on a deserted island. He has two blue pills and two red pills. He must take exactly one red pill and one blue pill to survive. How does he do it?

Solution : Break each of the pills in half, as you do this pop one half in your mouth and discard the other half.

2.8 Game Theory - Solution

Question : Player A invites player B to play the following game: A picks an integer n between 1 and 100, and writes it on a paper. B tries to guess n. If he succeeds, he receives n dollars. What is the fair price of the game, and what should be the strategy of B?

Solution : A tempting strategy for player A is to pick the lowest number 1 and to be guaranteed to lose at most 1. The expected loss of the winning strategy will have to be lower than 1. The key in this type of question is to observe that both players have access to the same amount of information. Therefore player B will guess the optimal strategy of A and take full advantage of it. We denote $p(i)$ the discrete probability distribution that A decides to use i for his choice. When player B picks a number he has an expected gain equal to $g_i = p(i).i$. Remember that player B will guess the probability distribution p, he will try to maximize his gain and player A will minimize the quantity

$$M = \max_{i \in [1,100]} g_i$$

This maximum is minimized when all the elements are equal $p(i).i = \lambda$. We find λ using the probability distribution properties

$$\sum_1^{100} p(i) = \sum_1^{100} \frac{\lambda}{i} = 1$$

$$\lambda = \frac{1}{\sum_1^{100} \frac{1}{i}} \approx \frac{1}{1 + \ln(n)}$$

therefore player A will pick the number i with a probability $p(i) = \frac{\lambda}{i}$. The expected loss (gain) for A (B) is $G = p(i).i = \lambda$. The numerical application with 100 numbers gives $G \approx 0.18$.

2.9 Stable Equilibrium I - Solution

Question : N tigers circle around an antelope. If a tiger eats an antelope or another tiger, it falls asleep and it becomes a potential meal for the remaining tigers. Tigers will eat if it does not endanger their life. The antelope keeps grazing quietly. Why?

Solution : In this classic type of question an unexpected equilibrium appears in a system. The best way to understand it is to start with a small number of tigers.

- 1 tiger: the tiger clearly eats the antelope, he does not need to worry about sleeping after the meal.

- 2 tigers: if a tiger eats the antelope he gets eaten by the other tiger. Tigers know that and decide to stay still. The system with 2 tigers is a stable system.

- 3 tigers: tigers have read this book and know that the 2 tigers system is stable, one of them eats the antelope, falls asleep and becomes the pray in a stable 2 tigers system.

- 4 tigers: tigers know that the 3 tigers system is unstable and prefer not to eat the antelope, the 4 tigers system is stable.

It appears that systems with an even number of tigers are stable. The antelope is relaxed because she has counted the tigers and found an even number.

2.10 Stable Equilibrium II - Solution

Question : 100 silent monks live in a monastery with no mirrors or reflective surfaces and one important rule: no red eyes! If a monk discovers he has red eyes he commits suicide at midnight. They live happily together in peace until a tourist visiting the monastery says "at least one of you has red eyes!". What happens next?

Solution : This is a different version of a classic type of equilibrium puzzles. We start with the cases with a low number of red eyed monks (RE group).

- Zero RE monk and the tourist lied to them: on the first day all the monks think that they are RE because they cannot see anyone else in RE. At midnight they all commit suicide. This bad prank should not happen because the tourist is assumed to tell the truth.

- 1 RE monk: the RE monk cannot see anyone else in RE and commits suicide at midnight.

- 2 RE monks: RE monks think on the first day that there is only one RE monk and they can see it. But no one commits suicide on the first night. At that point they realize that they are in a system with 2 RE and they both commit suicide on the second night.

- 3 RE monks: RE monks think that they are in a system with 2 RE, but no one commits suicide on the second night, they realize that it is a 3 RE system and they all commit suicide on the third night.

The pattern is clear, in conclusion in a system with j RE monks, all the RE monks commit suicide on the jth night.

2.11 Need For Speed - Solution

Question : A car travels 100km in 1 hour. Show that, at some point, its speed was exactly 100km/h.

Solution : This is a recurrent type of question based on the continuity of a function or its derivative. We denote $x(t)$ the position of the car at time t. $x(0) = 0$, $x(1) = 100$ and the mean value theorem (see page 51) proves that

$$\exists c \in [0,1] : x'(c) = \frac{x(1) - x(0)}{1} = 100$$

Alternatively we can use the intermediate value theorem (see page 51), if the average speed is 100, the speed cannot always be higher than 100, and it cannot always be lower than 100. There exists therefore a moment t_h where the speed is greater or equal to 100 and a moment t_l where the speed is lower or equal 100. Therefore $x'(t_h) \geq 100$ and $x'(t_l) \leq 100$ and $\exists c : x'(c) = 100$.

2.12 Russian Coin I - Solution

Question : Three players A, B and C sit around a table. They have a fair coin which gives heads or tails with a probability $\frac{1}{2}$. Player A tosses the coin, if he gets heads he wins, and the game is over. Otherwise he gives the coin to B, who is sitting at his right hand side. If B gets heads he wins, otherwise he gives the coin to C etc... What is the probability for each player to win the game?

Solution : There is an elegant way to solve this problem based on the symmetry of the players position. We define p_A (resp. p_B, p_C) the probability that player A (resp. B, C) wins the game and p as follows

$$p = P\{\text{Player who starts wins the game}\}$$

we see clearly that $p_A = p$. By symmetry, if player A misses his first toss, player B finds himself in the position of starting the same game. Therefore

$$p_B = P\{\text{A misses the first toss} \cap \text{Player who starts wins the game}\} = \frac{p}{2}$$

$$p_C = \frac{p}{4}$$

Also the probability that no one wins is zero

$$P\{\text{No one wins}\} = \lim_{\infty} \frac{1}{2}^n = 0$$

and

$$p_A + p_B + p_C = p + \frac{p}{2} + \frac{p}{4} = 1$$

$$p = \frac{4}{7} = p_A; \quad p_B = \frac{2}{7}; \quad p_C = \frac{1}{7}$$

The question can also be solved with series. We find that

$$p_A = \frac{1}{2} + \frac{1}{2} \cdot \frac{1}{2}^3 + \cdots + \frac{1}{2} \cdot \frac{1}{2}^{3i}$$

$$p_A = \frac{1}{2} \sum_{i=0}^{\infty} \frac{1}{8}^i = \frac{1}{2} \frac{1}{1-\frac{1}{8}} = \frac{4}{7}$$

(see Taylor series in page 52)

2.13 Russian Coin II - Solution

Question : Three players A, B and C sit around a table. They have a strange coin which gives heads or tails with a probability $\frac{1}{4}$, and stays stuck on its side with a probability $\frac{1}{2}$. Player A tosses the coin, if he gets side he wins, and the game is over. Otherwise if A gets heads he gives the coin to B, who is sitting at his right hand side. If A gets tails he gives the coin to C, who is sitting at his left hand side. The next player restarts the same process. What is the probability for each player to win the game?

Solution : We denote $P(i|j)$ the probability of the player j winning a game started by the player i. The probability of no one winning is equal to zero

$$P\{\text{No one wins}\} = \lim_{\infty} \frac{1}{2}^n = 0$$

therefore

$$P(A|A) + P(B|A) + P(C|A) = 1$$

We can visualize the possible outcomes after A turn

A plays $\begin{cases} \text{A gets heads with a probability } \frac{1}{4}; \text{ B wins with a probability } P(B|B) \\ \text{A gets tails with a probability } \frac{1}{4}; \text{ B wins with a probability } P(B|C) \\ \text{A gets side with a probability } \frac{1}{2}; \text{ A wins} \end{cases}$

and
$$P(B|A) = \frac{1}{4}P(B|B) + \frac{1}{4}P(B|C)$$

and by symmetry $P(B|B) = P(A|A)$ and $P(B|C) = P(C|A)$ giving

$$P(B|A) = \frac{1}{4}P(A|A) + \frac{1}{4}P(C|A)$$

By symmetry we also have
$$P(B|A) = P(C|A)$$

so the system of equations is

$$P(A|A) + 2P(B|A) = 1$$

$$\frac{3}{4}P(B|A) = \frac{1}{4}P(A|A)$$

We find $P(A|A) = \frac{3}{5}$ and $P(B|A) = P(C|A) = \frac{1}{5}$.

2.14 Be My Guest - Solution

Question : N guests are queuing at the entrance to get seated at a wedding table. Every guest has an assigned seat number but the first guest to choose his seat is too drunk and takes a random seat. The remaining guests choose their seat according to the following rule:

- if their assigned seat is available they take it

- if their assigned seat is taken they choose randomly an available seat

What is the probability that the last person gets his assigned seat?

Solution : Let us say the drunk person's seat is the number 1 and the last person's assigned seat is n. If at any moment a displaced person randomly chooses the seat number n, then the last person cannot get his assigned seat. But the critical remark is that, if at any time a displaced person randomly chooses the seat number 1, then the last person get his assigned seat. The chain of displaced guests is a cyclical permutation of the chain of assigned seats and choosing the seat 1 closes the cycle.

Guest	k	1	i	j
Seat	1	i	j	k

Therefore, as long as the seats 1 and n are available, the entering guest k has the following options

$$\begin{cases} p = \frac{1}{k} \text{ to pick 1, the last person is not displaced} \\ p = \frac{1}{k} \text{ to pick n, the last person is displaced} \\ p = \frac{k-2}{k} \text{ to pick another seat, the choice between 1 and n is postponed} \end{cases}$$

we can ignore how often the choice between 1 and n is postponed, when it finally happens the probabilities to choose 1 or n are equal. The probability that the last person gets his assigned seat is $\frac{1}{2}$.

2.15 4 Coins, 1 Table - Solution

Question : 4 coins are placed at the corners of a rotating table and the player is blindfolded. At every turn, the player can flip as many coins as he wants, and ask the game master if the coins are all showing heads. If they are all heads, the players wins, otherwise the game master can arbitrarily rotate the table before the next turn. Is there a winning strategy for the player?

Solution : We use the notation [h,h,h,t] for the current coins position, where h stands for heads and t for tails. We group the coins positions in classes which are stable by cyclical permutation. That means for example that [t,h,h,h], [h,t,h,h], [h,h,t,h] are grouped in the same class. Each time the player asks the game master if the current position is a winning position he can also flip all the coins and test the complementary position too. Therefore we can include the complementary sets in the classes, which means that [t,h,h,h], [h,t,t,t], [h,t,h,h], [t,h,t,t] etc... are in a same class.

We use the notation [f,o,o,o] to indicate which coins are flipped by the player, f stands for flipped and o indicates that the coins are not flipped. Similarly we group the player moves in classes which are stable by cyclical permutation.

$$\text{Position Classes} \begin{cases} p_1 : [h,h,h,h] \\ p_2 : [t,h,h,h] \\ p_3 : [t,t,h,h] \\ p_4 : [t,h,t,h] \end{cases} \quad \text{Transition Classes} \begin{cases} t_1 : [f,f,f,f] \\ t_2 : [f,o,o,o] \\ t_3 : [f,f,o,o] \\ t_4 : [f,o,f,o] \end{cases}$$

The transition t_1 is used at every step to check the complementary position. The position class p_1 is a winning position. We draw the following transition diagram

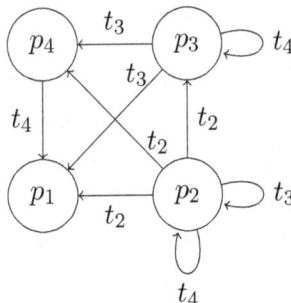

We notice that the transition t_4 applied on p_4 always leads to p_1. We can also see that p_2 and p_3 are stable by t_4. Note that the diagram does not include some adverse transitions, for example t_2 applied to p_4 sends back to p_2. But we can find a winning strategy with the information available in the diagram

- We ask if the starting position is a winning one. If not we are not in p_1

- We start by applying t_4. If we land in a winning position then we were in p_4. If not, we are either in p_2 or p_3.

- We apply now t_3 and then t_4. If we land in a winning position (after applying t_3 or after applying t_4) we can confirm we were in p_3 otherwise it means we were in p_2

- We know now that we are in p_2. We apply t_2, t_3 and t_4 to win.

The winning algorithm including the complementary check is therefore

$$t_4, t_1, t_3, t_1, t_4, t_1, t_2, t_1, t_3, t_1, t_4, t_1$$

2.16 N Coins, 1 Table - Solution

Question : 2 players take turns placing coins on a large perfectly round table. Coins can not overlap and all the coin surface must be in contact with the table. The first player who can't place a coin loses. Is it better to play first and is there a winning strategy?

Solution : The table is round and the winning strategy in this game is based on the central symmetry of the table. The first player A places his coin at the exact center of the table. Every time B places a coin A can respond by placing his coin in the symmetric position. With this strategy B is forced to discover new areas and A is guaranteed to place a coin. Therefore B will eventually run out of space and A is certain to win.

2.17 Regression Mirror - Solution

Question : Suppose that X and Y are mean zero, unit variance random variables. If least squares regression (without intercept) of Y against X gives a slope of β (i.e. it minimises $\mathbb{E}[(Y-\beta X)^2]$), what is the slope of the regression of X against Y?

Solution : We know that the slope of the least squares regression of Y against X is
$$\beta = (X^T X)^{-1} X^T Y$$
In this case X and Y are mean zero, unit variance random variables. Therefore
$$X^T X = \mathrm{Var}(X) = \sigma_X^2 = 1$$
$$X^T Y = \mathrm{Cov}(X, Y)$$
$$\beta = \mathrm{Cov}(X, Y)$$
and
$$\beta = \gamma$$
where γ is the slope of the regression of X against Y.

2.18 Bayesian Kids - Solution

Question : I meet someone with 2 children, and I learn that one of the children is a boy. What's the probability that the other child is also a boy? What if one of the children is a boy born on a Tuesday?

Solution : This is a classic question where the intuitive answer $\left(\frac{1}{2}\right)$ is wrong. To get it right, we write all the possible configurations, they all have a probability equal to $\frac{1}{4}$
$$BB, GG, BG, GB$$
The given information restrains the universe to BG, GB, BB and it appears that the probability that the other child is a boy is actually $\frac{1}{3}$.

Although the provided information seemed symmetric, it actually separated the universe into two blocks where the desired property is skewed. Hence the paradox. Let us see the second case, one of the boys is born on a Tuesday. We create a new property, X_T is a child born a Tuesday and X_O is a child born on another day of the week. Clearly for any child the probability to be born on a Tuesday is $\frac{1}{7}$ and the probability to have a boy born on a Tuesday is $\frac{1}{14}$. The configurations which are compatible with the provided information are now
$$G_T B_T, G_O B_T, B_T G_T, B_T G_O, B_T B_T, B_T B_O, B_O B_T$$

These cases correspond to the event $A =${At least one child is a boy born on a Tuesday} and we denote $A^c =${No child is a boy born on a Tuesday}. We have

$$P(A^c) = \left(\frac{13}{14}\right)^2$$

$$P(A) = 1 - P(A^c) = \frac{27}{196}$$

and the cases where the other child is also a boy are

$$B_T B_T, B_T B_O, B_O B_T$$

Let us call C this subset of A.

$$P(C) = \frac{1}{196} + \frac{6}{196} + \frac{6}{196} = \frac{13}{196}$$

and the desired probability is

$$\frac{P(C)}{P(A)} = \frac{13}{27}$$

2.19 Last Digit - Solution

Question : Consider all 100 digit numbers, i.e. those between 0 to $(10^{100} - 1)$, inclusive. For each number, take the product of non-zero digits (treat the product of digits of 0 as 1), and sum across all the numbers. What's the last digit?

Solution : We consider the problem between 0 and $(10^n - 1)$, let I_n be the sum of the product of the digits in the case n. For $n = 1$

$$I_1 = 1 + \sum_{i=1}^{9} i = 46$$

There is an induction relationship between I_n and I_{n-1}, when we fix a digit in the case $n + 1$ we are left with the sum of the product of the digits in $n - 1$

$$I_n = 1.I_{n-1} + \sum_{i=1}^{9} i.I_{n-1}$$

Therefore

$$I_n = 46 I_{n-1}$$

and

$$I_n = 46^n$$

To conclude, we notice that if two integers end with a 6, their product ends with a 6. The last digit is therefore a 6.

2.20 Repeated Contraction - Solution

Question : Let $R(n)$ be a random draw of integers between 0 and $n-1$ (inclusive). I repeatedly apply R, starting at 10^{100}. What's the expected number of repeated applications until I get zero?

Solution : Let F_n be the expected number or repeated applications when we start at n.
$$F_n = 1 + \frac{1}{n}(F_{n-1} + F_{n-2} + \cdots + F_0)$$
where $F_0 = 0$. And
$$F_{n-1} = 1 + \frac{1}{n-1}(F_{n-2} + F_{n-3} + \cdots + F_0)$$
Therefore
$$nF_n - (n-1)F_{n-1} = 1 + F_{n-1}$$
$$F_n = \frac{1}{n} + F_{n-1}$$
So $F_0 = 0$ and when $n > 0$
$$F_n = \sum_{i=1}^{n} \frac{1}{i} \simeq \ln(n)$$
and for $N = 10^{100}$
$$F_N \simeq 100 \ln(10) \simeq 230$$

2.21 Domino's Pizza - Solution

Question : How many ways are there to tile dominos (with size 2 × 1) on a grid of 2 × n? How about on a grid of 3 × 2n?

Solution : It is clear that in the case 2 x 1 the answer is 1, and in the case 2 x 2 the answer is 2.

The case 2 x n can be separated into 2 sub-cases. The tiling can either end with a horizontal or a vertical domino

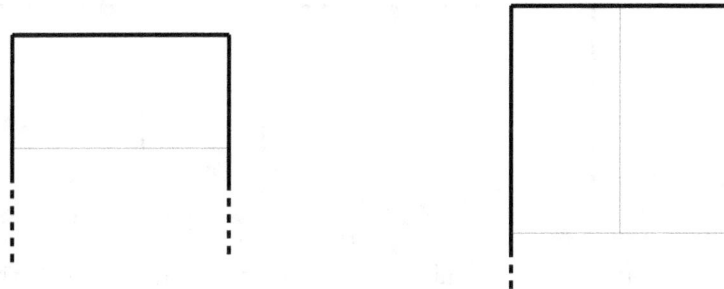

We denote I_n the total number of tiling for 2 x n. H_n and V_n the number of tiling ending respectively with a horizontal and a vertical domino. We have the following induction formula

$$I_n = H_n + V_n = I_{n-1} + I_{n-2}$$

We find that $I_n = \text{Fib}_{n+1}$ is a shifted Fibonacci sequence. The sequence equation is

$$r^2 - r - 1 = 0$$

with roots

$$r_1 = \frac{1+\sqrt{5}}{2}, \quad r_2 = \frac{1-\sqrt{5}}{2}$$

the Fibonacci sequence verifies

$$\text{Fib}_n = \frac{1}{\sqrt{5}}r_1^n - \frac{1}{\sqrt{5}}r_2^n$$

and therefore

$$I_n = \frac{1}{\sqrt{5}}r_1^{n+1} - \frac{1}{\sqrt{5}}r_2^{n+1}$$

To solve the configuration 3 x 2n, we start with 3 x 2. There are 3 possible tiling options

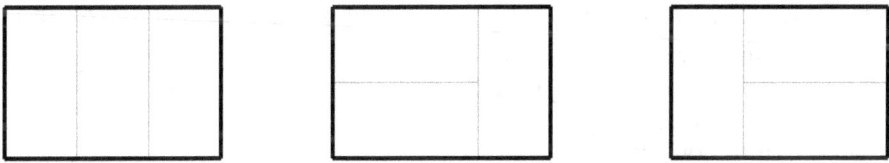

Actually, in the general case 3 x 2n, the tiling has to end with one of these 3 configurations

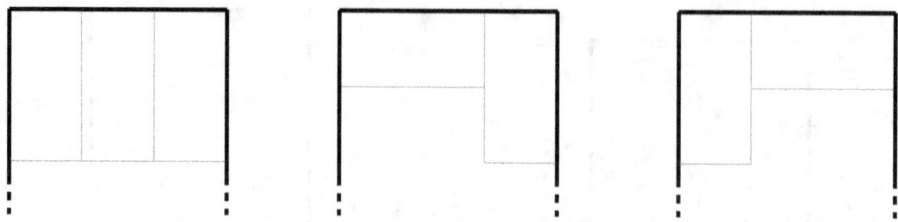

We denote C_n, L_n and R_n the number of configurations ending respectively with 3 vertical dominos (Center), only 1 vertical domino on the left and only 1 vertical domino on the right. We denote J_n the total number of configurations in the case 3 x 2n. By symmetry $L_n = R_n$ and

$$J_n = C_n + L_n + R_n = C_n + 2R_n$$

Clearly, every 3 x $2(n-1)$ admissible configuration can be extended with a center pattern to produce an admissible 3 x 2n pattern. therefore

$$C_n = J_{n-1}$$

However, we can construct a non centered (Left or Right) configuration with 2 methods: with or without overlapping. The figure below illustrates the Right case

We observe that every 3 x $2(n-1)$ admissible configuration can be extended with a Right pattern to produce an admissible 3 x 2n Right pattern without overlapping. But also, every 3 x $2(n-1)$ admissible Right configuration can be modified and extended with a Right pattern to produce an admissible 3 x 2n Right pattern with overlapping. Therefore

$$R_n = J_{n-1} + R_{n-1}$$

and the system of equations

$$\begin{cases} J_n = J_{n-1} + 2R_n \\ R_n = J_{n-1} + R_{n-1} \end{cases}$$

We substitute the second equation in the first equation

$$J_n = J_{n-1} + 2J_{n-2} + 2R_{n-2}$$

and by repeating

$$J_n = J_{n-1} + 2J_{n-2} + 2J_{n-3} + \cdots + 2J_1 + 2R_1$$

which is also true if we start from $n-1$

$$J_{n-1} = J_{n-2} + 2J_{n-3} + 2J_{n-4} + \cdots + 2J_1 + 2R_1$$

by subtracting the two equations

$$J_n - J_{n-1} = J_{n-1} + J_{n-2}$$

$$J_n = 2J_{n-1} + J_{n-2}$$

This is a recursive sequence of order 2 with equation

$$r^2 - 2r - 1 = 0$$

$$r_1 = 1 + \sqrt{2},\ r_2 = 1 - \sqrt{2}$$

We solve

$$J_n = Ar_1^n + Br_2^n,\ J_1 = 3,\ J_2 = 11$$

and we find

$$J_n = \left(\frac{5 - \sqrt{2}}{2}\right)\left(1 + \sqrt{2}\right)^n + \left(\frac{5 + \sqrt{2}}{2}\right)\left(1 - \sqrt{2}\right)^n$$

2.22 Nash's Car - Solution

Question : A company has a competition to win a car. Each contestant needs to pick a positive integer. If there's at least one unique choice, the person who made the smallest unique choice wins the car. If there are no unique choices, the company keeps the car and there's no repeat of the competition. It turns out that there are only three contestants, and you're one of them. Everyone knows before picking their numbers that there are only three contestants. How should you make your choice?

Solution : This question is a variation of the Game Theory question (see 1.8). A tempting strategy for all players is to choose the smallest number. But by doing so, they increase their chance to choose the same number as a competitor.

The key in this type of question is to observe that all players have access to the same amount of information. Therefore all players will guess the optimal strategy and take full advantage of it. We denote $p(i)$ the discrete probability distribution for every player to pick the number i for his choice. He wins the car if both competitors chose a larger number, or if if both competitors chose the same smaller number. His expected gain is

$$\mathbb{E}\left(\text{winnings} \mid \text{player picks } i\right) = E(i) = X \left(\sum_{j=1}^{i-1} p(j)^2 + \left(\sum_{j=i+1}^{\infty} p(j) \right)^2 \right) \quad (3)$$

where X is the price of the car. The optimal distribution should make the expected winnings independent of the chosen number. We search for a solution of the form

$$p(i) = Aq^i$$

where $q < 1$ and

$$A = \frac{1}{\sum_{i=1}^{\infty} q^i} = \frac{1-q}{q}$$

We inject it in equation (3)

$$E(i) = XA^2 \left(\sum_{j=1}^{i-1} q^{2j} + \left(\sum_{j=i+1}^{\infty} q^j \right)^2 \right)$$

$$E(i) = XA^2 \left(q^2 \frac{1-q^{2i-2}}{1-q^2} + \left(q^{i+1} \frac{1}{1-q} \right)^2 \right)$$

$$E(i) = XA^2 \left(\frac{(q^2 - q^{2i})(1-q) + q^{2i+2}(1+q)}{(1-q)^2(1+q)} \right)$$

$$E(i) = XA^2 \left(\frac{1-q}{1+q} + q^{2i+1} \frac{q^3 + q^2 + q - 1}{1+q} \right)$$

This quantity is constant when q is the only real root of

$$(q^3 + q^2 + q - 1) = 0$$

$$q_0 \simeq 0.5437$$

And we find that the optimal strategy is to pick the positive integer i with a probability

$$p(i) = (1-q_0)q_0^{i-1}$$

2.23 Correlation Impossible I - Solution

Question : If X, Y and Z are three random variables such that X and Y have a correlation of 0.9, and Y and Z have correlation of 0.8, what are the minimum and maximum correlation that X and Z can have?

Solution : This is a classic question, we want to find the values of $\text{Corr}(X, Z)$ for which the following matrix is a valid correlation matrix

$$\begin{pmatrix} 1 & 0.9 & x \\ 0.9 & 1 & 0.8 \\ x & 0.8 & 1 \end{pmatrix}$$

It is a valid correlation matrix if and only if it is positive semidefinite. This is equivalent to having no negative eigenvalue (see page 52). We derive the matrix characteristic polynomial

$$P(Y) = \begin{bmatrix} 1-Y & 0.9 & x \\ 0.9 & 1-Y & 0.8 \\ x & 0.8 & 1-Y \end{bmatrix}$$

$$P(Y) = (1-Y)^3 + 1.44x - 1.45 - x^2$$

The roots of P verify

$$(1-Y)^3 = x^2 - 1.44x + 1.45$$

and $Y \geq 0$ is equivalent to $(1-Y)^3 \leq 0$. Therefore our condition is

$$Q(x) = x^2 - 1.44x + 1.45 \leq 0$$

The roots of Q are $a \simeq 0.458$ and $b \simeq 0.982$, and the acceptable interval for the correlation between X and Z is

$$I = [a, b]$$

2.24 Correlation Impossible II - Solution

Question : If $X_1, X_2...X_n$ are n random variables such that

$$\text{Corr}(X_i, X_j) = \rho \text{ for } i \neq j$$

what are the minimum and maximum values that ρ can have?

Solution : We consider the matrix A

$$A = \begin{pmatrix} 1 & \rho & \cdots & \rho \\ \rho & 1 & \ddots & \vdots \\ \vdots & \ddots & \ddots & \rho \\ \rho & \cdots & \rho & 1 \end{pmatrix}$$

It is a valid correlation matrix if and only if it is positive semidefinite. This is equivalent to having no negative eigenvalue. For large matrices with patterns the eigenvalues can often be found visually. We notice that

$$A \begin{pmatrix} 1 \\ 1 \\ \vdots \\ 1 \end{pmatrix} = AX = \begin{pmatrix} 1 + (n-1)\rho \\ 1 + (n-1)\rho \\ \vdots \\ 1 + (n-1)\rho \end{pmatrix} = (1 + (n-1)\rho) X$$

We notice another pattern when we subtract two lines of A. Let $Y_i, i > 1$ the family of vectors defined as

$$\begin{cases} Y_i[1] = 1 \\ Y_i[i] = -1 \\ Y_i[j] = 0 \text{ otherwise} \end{cases}$$

We have

$$AY_i = (1 - \rho)Y_i$$

The family of vectors $(X, Y_2 \ldots Y_n)$ is therefore an independent family of n eigenvectors and the eigenvalues of A are $((1 - \rho), 1 + (n-1)\rho)$. Hence the valid values of ρ are

$$1 + (n-1)\rho \geq 0$$

$$\rho \geq -\frac{1}{n-1}$$

2.25 The Dark Side of the Die - Solution

Question : How many times do I have to roll a die until all six sides appear?

Solution : The trick for this question is to consider the expected number of rolls to see a new side. We denote respectively P_i (E_i) the probability (the expected number of rolls) to see a new side when we have already seen i sides. It is relatively straightforward that

$$E_i = \frac{1}{P_i}$$

When we start, the probability to see a new side at the next roll is 1. So $P_0 = E_0 = 1$. When we have seen i sides, we have $P_i = \frac{6-i}{6}$. Therefore the total expected number of rolls to see all sides E is

$$E = \sum_{0}^{5} E_i = \sum_{0}^{5} \frac{1}{P_i} = \sum_{0}^{5} \frac{6}{6-i} = \frac{6}{6} + \frac{6}{5} \cdots + \frac{6}{1} = 14.7$$

P.S. We denote E the expected number of attempts to make an event of probability p happen. We have

$$E = \sum_{i=1}^{+\infty} ip(1-p)^{i-1}$$

We identify the Taylor development of $\frac{1}{(1-x)^2}$ (see page 52) and

$$E = \frac{p}{(1-(1-p))^2} = \frac{1}{p}$$

2.26 Bonus Day - Solution

Question : Five pirates P_i have 100 gold coins. They have to divide up the loot. In order of seniority (suppose pirate P_5 is most senior, P_1 is least senior), the most senior pirate proposes a distribution of the loot. They vote and if at least 50% accept the proposal, the loot is divided as proposed. Otherwise the most senior pirate is executed, and they start over again with the next senior pirate. Which solution does the most senior pirate propose? Assume they are very intelligent and extremely greedy (and that they would prefer not to die).

Solution : As it is often the case in this type of questions, we need to work it out backwards starting with smaller systems.

- If there was only 1 pirate, he would take 100 coins.

- In a system with 2 pirates, the most senior would also take 100 coins as he is guaranteed to win the vote.

- In a system with 3 pirates, P_3 cannot have 100 coins because he cannot win the vote alone. But he could take 99 coins if he gives 1 coin to P_1. P_1 would be making more than what he could expect from smaller systems and would vote for the proposal.

- In a system with 4 pirates, the most senior pirate must convince one pirate to accept the proposal. We know that P_2 gets nothing in a 3 pirates system. Therefore P_4 can take 99 coins, give 1 coin to P_2 and win the vote. With one coin, P_2 has the opportunity to lock more than his expected gain in a 3 pirates system and he would accept the proposal.

- In a system with 5 pirates, P_5 needs to convince 2 pirates to vote with him. He will pick the pirates who would get nothing in a 4 pirates system: P_3 and P_1.

The most senior pirate should take 98 coins, give 1 coin to P_3 and 1 coin to P_1.

2.27 Secret Polynomial - Solution

Question : We consider a polynomial $P(x)$ which all coefficients are positive ($a_i \geq 0$). The polynomial is in a black box and we can only retrieve its value in given points. In how many points do we need to value the polynomial in order to find the values of all the coefficients?

Solution : The answer is 2. Let $P(x) = \sum_{i=0}^{n} a_i x^i$ be the polynomial, we start by taking the value of $P(1)$. $P(1)$ is greater than any coefficient a_i because all the coefficients are positive. We denote $M = P(1) + k$ where $k > 0$. Now we take the value of $P(M) = \sum_{i=0}^{n} a_i M^i$. The coefficients a_i are therefore the coefficients of the decomposition of $P(M)$ in base M and can now be found with a sequence of euclidean divisions.

2.28 Drunk Mutant Ninja Ant - Solution

Question : An ant starts a walk from a cube vertex, it walks on the edges and at every vertex it chooses to walk one of the available edges (including the edge it came from) with an equal probability. How many edges will the ant cross in average to come back to the starting point?

Solution : We assign numbers to the cube vertices and we denote 1 the vertex from which the ant starts its journey

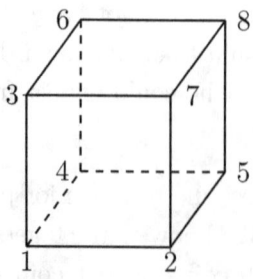

We can divide the vertices into groups or layers with interesting properties. We denote $L_1 = \{1\}$, $L_2 = \{2, 3, 4\}$, $L_3 = \{5, 6, 7\}$ and $L_4 = \{8\}$. We notice that the ant always has to change layer when it travels through an edge. We can draw the following transition diagram.

$$L_1 \xrightarrow[p=1]{p=\frac{1}{3}} L_2 \xrightarrow[p=\frac{2}{3}]{p=\frac{2}{3}} L_3 \xrightarrow[p=1]{p=\frac{1}{3}} L_4$$

We denote now N_i the average number of edges the ant needs to travel to come back to the vertex 1 given that it starts from the layer i. We have the system

$$\begin{cases} N_1 = 1 + N_2 \\ N_2 = \frac{1}{3} + \frac{2}{3}(1 + N_3) \\ N_3 = \frac{2}{3}(1 + N_2) + \frac{1}{3}(1 + N_4) \\ N_4 = 1 + N_3 \end{cases}$$

We can eliminate N_1 and N_4

$$\begin{cases} N_2 = \frac{1}{3} + \frac{2}{3}(1 + N_3) \\ N_3 = \frac{2}{3}(1 + N_2) + \frac{1}{3}(2 + N_3) \end{cases}$$

we solve to find

$$\begin{cases} N_1 = 8 \\ N_2 = 7 \\ N_3 = 9 \\ N_4 = 10 \end{cases}$$

And the answer is 8.

2.29 Dog Day Afternoon - Solution

Question : You are standing at the centre of a circular field of radius R. The field has a low wire fence around it. Attached to the wire fence (and restricted to running around the perimeter) is a large, sharp-fanged, hungry dog. You can run at speed v, while the dog can run four times as fast. What is your running strategy to escape the field?

Solution : This is a surprisingly popular question in interviews. The key is to keep things simple. If you run to the fence from the center and away from the dog, it would leave you running a distance R and the dog running πR this would take you a time R/v and him $\pi R/(4v)$ so clearly he would get there before you as

$$\frac{R}{v} > \frac{\pi R}{4v}$$

First let us find the smallest distance from the center R_1 from which running away from the dog is a winning strategy. The dog needs to run πR to catch up and R_1 verifies

$$\frac{R - R_1}{v} = \frac{\pi R}{4v}$$

and

$$R_1 = R - \frac{\pi R}{4} \simeq 0.215 R$$

Now when we are on a smaller circle we can have a higher angular speed than the dog. The other interesting radius is the longest distance from the center R_2 for which we have a higher angular speed than the dog. We compare the times to run a full circle to find R_2, R_2 verifies

$$\frac{2\pi R}{4v} = \frac{2\pi R_2}{v}$$

and

$$R_2 = \frac{R}{4} = 0.25 R$$

So the strategy is clear: walk to a distance R_3 from the center such that $R_3 \in]R_1, R_2[$. Then run along the circle of radius R_3 until the dog is diametrically opposed to you, it is possible because $R_3 < R_2$. Then run towards the fence, you will make it before the dog because $R_3 > R_1$.

2.30 Exp Pi Red - Solution

Question : Is $\pi^e > e^\pi$?

Solution : We want to compare

$$e \ln(\pi) \overset{?}{>} \pi \ln(e)$$

equivalent to

$$\frac{\ln(\pi)}{\pi} \overset{?}{>} \frac{\ln(e)}{e}$$

So we are interested in the function

$$f(x) = \frac{\ln(x)}{x}$$

$f(1) = 0$ and $\lim_{+\infty} f = 0$, the maximum of f in $[1, +\infty]$ satisfies

$$f'(x) = \frac{1 - \ln(x)}{x^2} = 0$$

so we find that f is at its highest value at $x = e$ and

$$\frac{\ln(\pi)}{\pi} < \frac{\ln(e)}{e}$$

and

$$e \ln(\pi) < \pi \ln(e)$$

2.31 Brilliant - Solution

Question : You're in a room with three light switches, each of which controls one of three light bulbs in the next room. You need to determine which switch controls which bulb. All lights are off to begin, and you can't see into one room from the other. You can inspect the other room only once. How can you find out which switches are connected to which bulbs?

Solution : We denote the switches 1, 2 and 3. Switch on switches 1 and 2, wait a moment and switch off number 2. Enter the room. Whichever bulb is on is wired to switch 1, whichever is off and hot is wired to switch number 2, and the third is wired to switch 3.

2.32 Lognormal Expectation - Solution

Question : Calculate $\mathbb{E}(\exp(X))$ when is X is a normally distributed random variable

$$X \sim \mathcal{N}\left(\mu, \sigma^2\right)$$

Solution : Apply the definition of an expectation (see page 51) to the random variable $Y = \exp(X)$ and use the probability density of a normal distribution (see page 51)

$$\mathbb{E}(\exp(X)) = \int_{-\infty}^{\infty} \exp(x) \frac{1}{\sqrt{2\pi}\sigma} \exp\left(-\frac{(x-\mu)^2}{2\sigma^2}\right) dx$$

$$\mathbb{E}(\exp(X)) = \frac{1}{\sqrt{2\pi}\sigma} \int_{-\infty}^{\infty} \exp\left(\frac{2\sigma^2 x - x^2 + 2\mu x - \mu^2}{2\sigma^2}\right) dx$$

$$\mathbb{E}\left(\exp(X)\right) = \frac{1}{\sqrt{2\pi}\sigma} \int_{-\infty}^{\infty} \exp\left(\frac{-(x-(\sigma^2+\mu))^2}{2\sigma^2} + \frac{\sigma^4 + 2\mu\sigma^2}{2\sigma^2}\right) dx$$

$$\mathbb{E}\left(\exp(X)\right) = \exp\left(\frac{\sigma^2}{2} + \mu\right) \int_{-\infty}^{\infty} \frac{1}{\sqrt{2\pi}\sigma} \exp\left(\frac{-(x-(\sigma^2+\mu))^2}{2\sigma^2}\right) dx$$

We identify the integral of the density of a normal distribution, equal to 1 and

$$\mathbb{E}\left(\exp(X)\right) = \exp\left(\frac{\sigma^2}{2} + \mu\right)$$

2.33 Cumulative Brownian - Solution

Question : Calculate $\mathbb{E}(\Phi(B_t))$ where B_t a brownian motion and Φ the standard normal cumulative distribution.

Solution : The elegant solution for this problem is based on the symmetry of the brownian motion (see page 53):

$$\mathbb{E}\left(\Phi\left(W_t\right)\right) = \mathbb{E}\left(\Phi\left(-W_t\right)\right) = \mathbb{E}\left(1 - \Phi\left(W_t\right)\right) = 1 - \mathbb{E}\left(\Phi\left(W_t\right)\right)$$

and we get

$$\mathbb{E}\left(\Phi\left(W_t\right)\right) = \frac{1}{2}$$

The result can also be proved using integrals

$$\mathbb{E}\left(\Phi\left(W_t\right)\right) = \int_{-\infty}^{+\infty} \left(\int_{-\infty}^{x} \frac{1}{\sqrt{2\pi}} \exp\left(\frac{-u^2}{2}\right) du\right) \frac{1}{\sqrt{2\pi t}} \exp\left(\frac{-x^2}{2t}\right) dt$$

$$\mathbb{E}\left(\Phi\left(W_t\right)\right) = \int_{-\infty}^{+\infty} \Phi(x)\Phi'(x) dx = \left[\frac{\Phi^2(x)}{2}\right]_{-\infty}^{+\infty} = \frac{1}{2}$$

2.34 Learn the Ropes - Solution

Question : You have two ropes coated in an oil to help them burn. Each rope will take exactly 1 hour to burn all the way through. However, the ropes do not burn at constant rates, there are spots where they burn a little faster and spots where they burn a little slower, but it always takes 1 hour to finish the job.

With a lighter to ignite the ropes, how can you measure exactly 90 minutes? And how can you measure 45 minutes?

Solution : In this classic question, the trick is to burn a rope at both ends at the same time. This way the rope burns in 30mn. That gives the answer for the first question, to measure 90mn you can burn the first rope at one end, and when the first rope is consumed (60mn) burn the second rope at both ends at the same time.

45mn is a bit trickier, somehow we need to produce 15mn. The trick here is to burn at time T the first rope at one end and the second rope at both ends. When the second rope is consumed, we are at (T+30) and the first rope is half consumed. At that time (T+30) we burn the first rope at its other end, doubling its burning speed and obtaining the required 15mn. This gives a total of 45 minutes.

2.35 Bayes Bias - Solution

Question : I have one fair coin and one biased two headed coin, and I put both in my pocket. I randomly choose one coin and flip it. It shows heads. What is the probability that the coin has Tails on the other side? What if you flip it n times and get heads n times?

Solution : This is an application of the Bayes theorem (see page 52) on conditional probabilities. The key here is to define the events correctly. We define

$$A = \{\text{I choose the biased coin from my pocket}\}$$

$$B = \{\text{I get heads when I flip the coin}\}$$

The Bayes theorem gives

$$P(A|B) = \frac{P(A \cap B)}{P(B)}$$

Note that if I pick the biased coin, I get heads with a probability 1. So

$$P(A \cap B) = P(A) = \frac{1}{2}$$

For $P(B)$ we need to treat both cases, with the fair coin and with the biased coin

$$P(B) = \frac{1}{2}.1 + \frac{1}{2}.\frac{1}{2} = \frac{3}{4}$$

and

$$P(A|B) = \frac{1}{2}.\frac{4}{3} = \frac{2}{3}$$

So if I get heads after one flip, I have a probability $\frac{2}{3}$ of having the biased coin, and a probability $\frac{1}{3}$ to have the fair coins with tails on the other side.

We can generalize for n flips, $P(A \cap B)$ is unchanged and $P(B)$ becomes

and
$$P(B) = \frac{1}{2} \cdot 1 + \frac{1}{2} \cdot \frac{1}{2^n} = \frac{1+2^n}{2^{n+1}}$$

$$P(A|B) = \frac{1}{2} \cdot \frac{2^{n+1}}{1+2^n} = \frac{2^n}{1+2^n}$$

So after n heads we have a probability $\frac{2^n}{1+2^n}$ to have the biased coin in hand and $\frac{1}{1+2^n}$ to have the fair coin.

2.36 Die Hard - Solution

Question : You've got to defuse a bomb by placing exactly 4 gallons of water on a sensor. You only have a 5 gallon jug and a 3 gallons jug on hand. How do you proceed?

Solution : This classic riddle, made famous in Die Hard 3, is still surprising popular in interviews.

- Fill the 5-jug up completely. There will be, of course, 5 gallons in the 5-jug. You must fill all the gallons up to the top, otherwise you don't actually know how much you have.

- Use the water from the 5-jug to fill up the 3-jug. You're left with 3 gallons in the 3-jug and 2 gallons in the 5-jug.

- Pour out the 3-gallon jug. You're left with nothing in the 3-jug and 2 gallons in the 5-jug.

- Transfer the water from the 5-jug to the three jug. You're left with 2 gallons in the 3-jug. And nothing in the 5-jug.

- Fill up the 5-jug completely. You now have 2 gallons in the 3-jug and 5 in the 5-jug. This means that there is 1 gallon of space left in the 3-jug.

- Use the water from the 5-jug to fill up the 3-jug. Fill up the last gallon of space in the 3-jug with the water from the 5-jug. This leaves you with 3 gallons in the 3-jug, and 4 gallons in the 5-jug.

2.37 Blind coins - Solution

Question : You have a 100 coins laying flat on a table, each with a head side and a tail side. 10 of them are heads up, 90 are tails up. You can't feel, see or in any other way find out which 10 are heads up. How can you split the coins into two piles so there are the same number of heads-up coins in each pile?

Solution : Pick 10 coins from the original 100 and put them in a separate pile P_n. The original pile P_o contains now m heads-up and $(90 - m)$ tails-up. The new pile P_n contains $(10 - m)$ heads-up and m tails-up. Then flip the 10 coins in P_n over. The two piles are now guaranteed to have the same number of heads m.

2.38 Russian roulette - Solution

Question : We are to play a version of Russian Roulette, the revolver is a standard six shooter but I will put two bullets in the gun in consecutive chambers. I spin the chambers, put the gun to my head pull the trigger and survive. I hand you the gun and give you a choice... You may put the gun straight to your head and pull the trigger, or you may re-spin the gun before you do the same.
What is your choice and why? How does this differ from the case with only one bullet?

Solution : We denote $\{C_1, ..., C_6\}$ the gun chambers, and we assume that the bullets are in C_1 and C_2. After the survival of the other player, the gun is in one of the following positions
$$C_4, C_5, C_6, C_1$$
where C_1 is the only loaded chamber. So before a re-spin the probability of survival is
$$p = \frac{3}{4}$$
If we decide to re-spin, the probability of survival is $\frac{4}{6} = \frac{2}{3}$. So it is better to not re-spin. If there was only one bullet, the possible positions before re-spin are
$$C_3, C_4, C_5, C_6, C_1$$
where C_1 is the only loaded chamber. So before a re-spin the probability of survival is
$$p = \frac{4}{5}$$
If we decide to re-spin, the probability of survival is $\frac{5}{6}$. In th one bullet case it is better to re-spin.

2.39 Rain Check - Solution

Question : You're about to get on a plane to Seattle. You want to know if it's raining there. You call 3 random friends who live there and ask each if it's raining.

Each friend has a 2/3 chance of telling you the truth and a 1/3 chance of messing with you by lying. All 3 friends tell you that "Yes" it is raining. What is the probability that it's actually raining in Seattle?

Solution : This is an application of Bayes Theorem. Let p_r be the prior probability for rain ($p_n \equiv 1 - p_r$). Then the probability of rain given 3 "yes" replies $\{y, y, y\}$ is

$$P(\text{rain} \mid \{y, y, y\}) = \frac{P(\text{rain} \cap \{y, y, y\})}{P(\{y, y, y\})}$$

$$= \frac{P(\{y, y, y\} \mid \text{rain}) \cdot p_r}{P(\{y, y, y\} \mid \text{rain}) \cdot p_r + P(\{y, y, y\} \mid \text{no rain}) \cdot p_n}$$

$$= \frac{(2/3)^3 \cdot p_r}{(2/3)^3 \cdot p_r + (1/3)^3 \cdot p_n} = \frac{p_r}{p_r + p_n/8}$$

We cannot go any further without assumption on the prior probability for rain. Usually the interviewer will ask you to use $p_r = 0.25$ and

$$P(\text{rain} \mid \{y, y, y\}) = \frac{8}{11}$$

2.40 No Time to Die - Solution

Question : A regular clock has an hour and minute hand. At 12 midnight the hands are exactly aligned. When is the next time they will exactly align or overlap?

Solution : This classic question can be solved very elegantly. Notice that, whenever the hands are exactly aligned, the waiting time before the next overlap is always the same. This can be proved because the position is symmetric by rotation. It can also be seen as the time needed for the fast hand to catch up on the slow hand, the angle to catch up and the speeds of rotation being constant. We define Δ the waiting time before the next overlap, We see that a 12 hours interval contains 11Δ. Therefore

$$\Delta = \frac{12}{11} hour = 1.091 hour = 65.45 mn = 1Hr5mn27.3s$$

The problem can also be solved with a simple angle calculation. After one lap, we denote α the fraction of a full circle where the hands meet

$$\frac{2\pi}{12} + \frac{2\pi}{12}\alpha = 2\pi\alpha$$

$$1 + \alpha = 12\alpha$$

$$\alpha = \frac{1}{11}$$

and

$$\Delta = 12\alpha hour = \frac{12}{11} hour = 1.091 hour = 65.45 mn = 1Hr 5mn 27.3s$$

Chapter 11

Math Cheatsheet

Math Cheatsheet

3.1 Normal Distribution

$$X \sim \mathcal{N}\left(\mu, \sigma^2\right)$$

$$f_X(x) = \frac{1}{\sqrt{2\pi}\sigma} \exp\left(-\frac{(x-\mu)^2}{2\sigma^2}\right)$$

3.2 Expectation

If X is a random variable with a probability density function of $f(x)$, then the expected value is defined as the Lebesgue integral

$$\mathrm{E}[X] = \int_{\mathbb{R}} x f(x) dx$$

3.3 Mean Value Theorem

Let $f : [a, b] \to \mathbb{R}$ be a continuous function on the closed interval $[a, b]$, and differentiable on the open interval (a, b), where $a < b$. Then there exists some c in (a, b) such that

$$f'(c) = \frac{f(b) - f(a)}{b - a}$$

3.4 Intermediate Value Theorem

Consider an interval $I = [a, b]$ of real numbers $\mathbb{R}$ and a continuous function $f : I \to \mathbb{R}$. Then, if u is a number between $f(a)$ and $f(b)$, that is,

$$\min(f(a), f(b)) < u < \max(f(a), f(b))$$

then there is a $c \in (a, b)$ such that

$$f(c) = u$$

3.5 Correlation

The population correlation coefficient $\rho_{X,Y}$ between two random variables X and Y with standard deviations σ_X and σ_Y is defined as

$$\rho_{X,Y} = \frac{\mathrm{cov}(X, Y)}{\sigma_X \sigma_Y}$$

3.6 Bayes' theorem

$$P(A|B) = \frac{P(B|A)P(A)}{P(B)} = \frac{P(A \cap B)}{P(B)}$$

where A and B are events and $P(B) \neq 0$

3.7 Eigenvalues

Let A be a linear transformation represented by a matrix A. If there is a vector $\mathbf{X} \in \mathbb{R}^n \neq \mathbf{0}$ such that

$$A\mathbf{X} = \lambda \mathbf{X}$$

for some scalar λ, then λ is called the eigenvalue of A with corresponding eigenvector $\mathbf{X}$.

3.8 Symmetric Matrices

Any symmetric matrix A $(A = A^T)$

- has only real eigenvalues
- is always diagonalizable
- has orthogonal eigenvectors

3.9 Semidefinite Positive Matrices

The symmetric matrix A is said positive semidefinite $(A \geq 0)$ if all its eigenvalues are non negative.

3.10 Useful Taylor Series

$$\frac{1}{1-x} = 1 + x + x^2 + x^3 + x^4 + \ldots$$

$$\left(\frac{1}{1-x}\right)' = \frac{1}{(1-x)^2} = 1 + 2x + 3x^2 + 4x^3 + 5x^4 + \ldots$$

$$e^x = 1 + x + \frac{x^2}{2!} + \frac{x^3}{3!} + \frac{x^4}{4!} + \ldots$$

$$\cos x = 1 - \frac{x^2}{2!} + \frac{x^4}{4!} - \frac{x^6}{6!} + \frac{x^8}{8!} - \ldots$$

$$\sin x = x - \frac{x^3}{3!} + \frac{x^5}{5!} - \frac{x^7}{7!} + \frac{x^9}{9!} - \ldots$$

3.11 Brownian Motion

A Brownian motion is a stochastic process $\{B_t\}_{t \geq 0+}$ with the following properties:

- $B_0 = 0$
- The function $t \to B_t$ is almost surely continuous in t
- The process $\{B_t\}_{t \geq 0}$ has stationary, independent increments
- The increment $B_{t+s} - B_s$ has the $\mathcal{N}(0, t)$ distribution

A Humble Request

Dear valued reader,

We are, Editions Ducourt, a small publishing company and without your support we would not exist.

Therefore we make a humble request - if you enjoy this book, please spare a few minutes to leave us a review on this book's Amazon product page.

<u>Each and every one</u> of your reviews is paramount to the success of the book as its visibility is impacted by the Amazon algorithm.

We are forever grateful for your support and we hope we have succeeded in providing you with a very special book.

Sincerely
Editions Ducourt

Index

Bayes, 52
Brownian Motion, 53

Correlation, 51

Eigenvalues, 52
Expectation, 51

Intermediate Value Theorem, 51

Lognormal, 9

Mean value theorem, 51

Normal Distribution, 51

Prime, 3

Ramsey, 3

Semidefinite Matrix, 52
Symmetric Matrices, 52

Taylor Series, 52

www.ingramcontent.com/pod-product-compliance
Lightning Source LLC
Chambersburg PA
CBHW081500220526
45466CB00008B/2730